DEFENDERS

Created and Produced by Firecrest Books Ltd
in association with John Francis/Bernard Thornton Artists

Copyright © 1999 Firecrest Books Ltd
and Copyright © 1999 John Francis/Bernard Thornton Artists

Published by Tangerine Press™, an imprint of Scholastic Inc.
555 Broadway, New York, NY 10012

 tangerine press

ISBN 0-439-15347-6

Printed and bound in Belgium
First printing December 1999

DEFENDERS

Bernard Stonehouse

Illustrated by
John Francis

TANGERINE PRESS™ and associated logo
and design are trademarks of Scholastic Inc.

For Sam

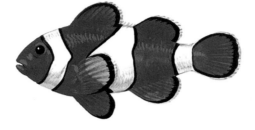

Art and Editorial Direction by
Peter Sackett

Designed by
Paul Richards, Designers & Partners

Edited by
Norman Barrett

Color separation by
Sang Choy International Pte. Ltd.
Singapore

Printed and bound by
Casterman, Belgium

— Contents —

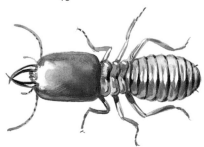

— Introduction STAYING ALIVE

People are impressed by the predators of the animal kingdom — the animals that eat other animals, take the initiative, make the attack, call the shots. As descendants of hunters, we ourselves are predators. We relate to other predators far more than to prey, which are the intended or actual victims. Predators we see as winners, prey as losers. We have sympathy for losers, but line ourselves up with winners, perhaps to share their luck, or to learn something from them. There is little luck in becoming someone else's dinner.

But prey are also winners. In the hard world of eat-and-be-eaten, any animal that survives to adulthood has already won several rounds. If it survives to reproduce and leave offspring, it has won the game. No one is a loser until they have lost, and prey that survive have won, on their own terms, by the tricks and strategies that we sum up as "defense." This book is about defenders, and how they hold on to whatever they value — homes, mates, offspring, and lives. We have much to learn from them, too.

Defenders hold out in dozens of ways, some obvious, others bizarre, all remarkable — a bird feigning injury, drawing predators away from its nest; an octopus disappearing in a puff of homemade ink; an albatross chick hurling its breakfast at marauding skuas; a fish blowing itself up into a ball of spines; a mother rhino fending off attackers; and musk oxen shielding their young within a wall of horns. These are remarkable enough but, no less remarkably, they are almost all inborn or preprogrammed responses. Mother rhino might have learned from experience to know how to protect her calf, but most of the others hardly understand what they are doing. Yet these simple ploys for defense save their own lives, and give their offspring a further chance of survival.

The pangolin is an inoffensive animal that has evolved an almost perfect defense — a "coat of mail" formed by overlapping horny scales. There are several species living in Africa and Asia, and they are similar to the armadillo (see page 10). When attacked they roll up into a ball that few enemies can penetrate.

Ringed plover FEIGNING INJURY

Ringed plovers are waders, or shorebirds, about 8 inches (20 cm) long. They breed in northern Europe, Greenland, Svalbard, Siberia, and northern Canada. In autumn and winter, huge flocks gather to feed on sandy beaches and estuaries of warmer regions.

In breeding plumage they are smart little birds, banded in black and white, with a black-tipped orange bill and brilliant orange feet. They nest inland, on sand or rocks alongside rivers, lakes, and reservoirs. Though many may nest within a few yards of one another, incubating birds are extraordinarily hard to see. Even harder to detect are the nests themselves. Mere scrapes in the rocks, they have no lining, and the three or four eggs and chicks match their background cannily.

In defense of their nests, ringed plovers are artists in deception. When a fox, weasel, or other predator approaches, the incubating bird leaps up and totters from the nest with wing trailing and tail spread, shrilling alarm calls as though seriously disabled. No predator can resist following. Eggs and nestlings remain safe, while the parent leads the predator on a dance well away from the nest.

Nine-banded armadillo PUZZLE BOX

Armadillos are widespread in South and Central America, and this species, the nine-banded armadillo, extends north into the southern United States. Coyotes are common throughout North America from Mexico to Alaska. Here, in a windswept desert corner, armadillo and coyote have met as predator and potential prey.

Armadillos burrow in the ground during the hot, sunny days, emerging in the evenings to feed. Bumbling through the grasslands and scrub like miniature tanks, they sniff out insects, lizards, snakes, birds and their eggs, small mammals, and carrion — remnants of animals killed by others and left to rot. Coyotes, too, eat practically anything, and by evening are ready for supper. They chase after lizards, birds, and mammals, including armadillos, which move surprisingly fast on their short legs. However, a coyote can run more quickly than they can. If the armadillo reaches a burrow, it is probably safe. If not, it stops and curls up tightly, "hinged" at the nine bands.

Now the coyote has a problem. Almost every square inch of an armadillo's back and flanks is armored with bony plates and cased in tough, leathery skin. Only the soft underbelly is vulnerable, and that is protected with dense fur. The coyote will soon conclude that a rolled-up armadillo leaves little for it to get its teeth into, so it wanders off in search of more cooperative prey.

The nine bands are strips of scales, with hinges between, that allow the armadillo to curl into a ball.

Porcupine fish THORNY PROBLEM

Porcupines are mammals that defend themselves with long, backward-pointing spines (see pages 42–43). Porcupine fish, like the one shown here, are covered with scales like other fish, but each scale ends in a sharp bony spine. There are several species living in tropical seas and estuaries. Most of them are 1–2 feet (30–60 cm) long. They feed on the seabed, crunching shellfish with their powerful jaws and flat teeth.

Left to themselves, porcupine fish look much like any other fish — rougher and spinier than most, but otherwise unremarkable. However, one that is alarmed stops swimming and pumps itself up with water like a balloon until it is round in shape. Now the spines on the tips of the scales stick out in all directions. A hungry shark that sees an ordinary-looking fish, and wants it for supper, is suddenly faced by a monstrous combination of basketball and pincushion, with the pins pointing outward rather than in. The porcupine fish treads water, turning on its own axis to keep the predator in sight. It stays that way until the danger passes, then lets out the water and returns to normal.

A porcupine fish hauled from the water in a fishing net can similarly fill itself with air and entangle itself thoroughly in the net — a good reason why fishermen prefer not to catch them.

A porcupine fish before inflating itself.

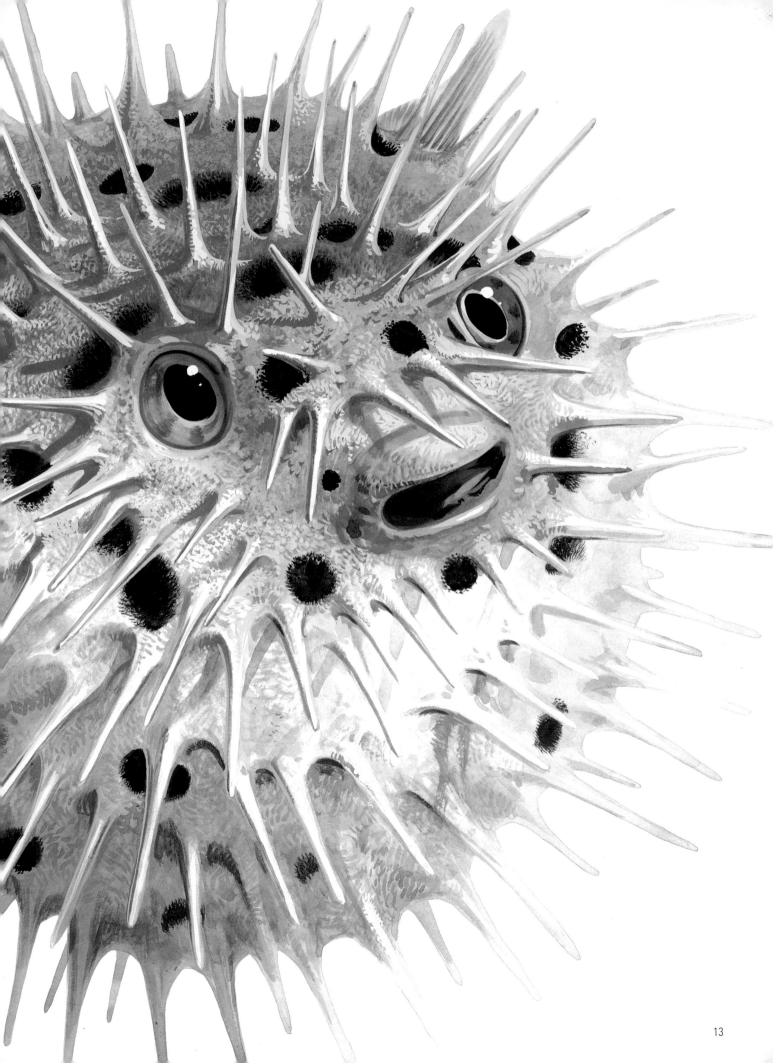

Black rhinoceros DRIVING OFF HYENAS

African black rhinos stand up to 5 feet (1.5 m) at the shoulder, and weigh as much as a small truck. Their favorite habitat is grassy plain and thornbush, with water close at hand, where they can browse on grass, leaves, and shoots, and enjoy a cooling evening wallow. Both males and females tend to be solitary. A group of two or three is probably a female with her calf and perhaps a half-grown young rhino seeking company and mutual protection.

With two horns on the nose, a leathery hide, short sight, and a shorter temper, this mother rhino is a formidable defender. Nothing is more likely to arouse her than a group of hungry hyenas. She alone is more than a match for them, but her week-old calf stands only 30 inches (75 cm) high, and weighs about 55 pounds (25 kg) — smaller than an adult spotted hyena. Five or six hyenas, working together, surround and confuse her.

The calf seems to know that safety lies alongside the fortress wall of its mother's body. As she turns and wheels to face the attackers, the calf turns with her, almost under her feet. The hyenas, daring but not foolhardy, keep outside the sweep of that nose and its two formidable horns. They eventually might have to give up. Well, it was always worth a try.

— Ferruginous rough-leg MOBBING

Mobbing is a powerful kind of defense that many birds use to protect their nests and young. In human terms, a mob is a crowd of people on the rampage — often weak folk using their collective strength against a more powerful adversary. This also describes the actions of birds that, faced with a predator, pool their resources to beat it. As individuals they have little strength, but together they can scare off a much larger foe.

Here is a coyote — a doglike predator that we have met before (pages 10–11) — looking for a meal. He has seen a ferruginous rough-leg on a nest, and probably remembers that an occupied nest is liable to contain eggs. Ferruginous rough-legs are large predatory hawks similar to European buzzards, with wingspans up to 5 feet (1.5 m) and possessing soaring flight. Though predators themselves, with a taste for rabbits, mice, and other small mammals, individually they are no match for a coyote.

So the breeding pair are mobbing. When the coyote first approached, the sitting bird sounded an alarm call that brought its partner screaming home. Now the birds alternately swoop over the coyote's head, screeching continuously, sometimes just touching, but keeping out of range of those fierce jaws. The coyote is distracted and bothered, as you or I would be, and may well give up.

Octopus DISAPPEARING ACT

About 150 different kinds of octopus swim the world's oceans. They are mollusks, distantly related to slugs and snails, but larger and more lively. The biggest are 30 feet (9 m) across, the smallest about 2 inches (5 cm). Eight arms, webbed at the base and lined with suckers, spread from a central body containing the vital organs. Octopuses vary widely in color. Some are banded, others blotched, and many change color as you watch.

This common octopus of Atlantic shores grows to about 10 feet (3 m), though the ones you see close to the shore are rarely more than 2 feet (60 cm) across. Active predators, they have sharp eyes and acute senses of smell and taste. Moving by a combination of jet propulsion and swimming strokes, the octopus creeps up on crabs, lobsters, shrimp, and fish, pouncing and enveloping them with those flexible arms, which draw the prey toward a sharp, parrot-like beak.

Predators one moment, they may, however, be prey the next. An octopus hunted by a hungry fish will first confuse it by changing color. Then it produces a cloud of specially secreted brown or black ink, of which a sacfull is always held in reserve. By the time the predator finds its way into the cloud, the octopus has darted off.

Giant horntailed sawfly SAWING AND DRILLING

If this yellow-and-black creature, almost 1 inch (2.5 cm) long, one day buzzes past your nose, your warning system will tell you "stinging insect." It is big, it buzzes, its coloring is vaguely wasplike, and its rear end carries a spike that looks remarkably like a stinger.

However, although it is known in Europe as a wood wasp, it is not a true wasp and it does not sting. It is a giant horntailed sawfly. Compared with a true wasp, its coloring and shape are wrong — it is too tubular and has no waist. The spike is used for drilling and egg-laying. The insect drills a succession of holes into trees or shrubs, laying an egg in each. The eggs turn into brown caterpillar-like larvae, which munch the leaves and shoots of the tree. The larvae spin webs and turn into cocoons, and from the cocoons emerge the new sawflies.

In this cycle of events, it is the fly that is most visible and vulnerable to predation from birds. So the sawfly gains protection by looking like a wasp. Humans avoid wasps because they sting, but that is not the point. Birds, we are told, avoid wasps because they taste nasty, so birds will avoid any other insects that look even vaguely like wasps.

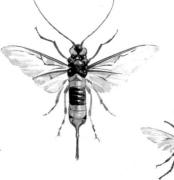

The giant horntailed sawfly, or wood wasp (left), compared with a true wasp.

Desert tortoise REPTILE IN A BOX

Of all the world's four-footed animals, land-living tortoises and aquatic turtles have the longest fossil history. They first appear recognizably in rocks that buried them 200 million years ago, and have trudged their way stolidly through the world ever since. Foxes by comparison are latecomers, first appearing only a few million years ago. Humans span an even shorter history.

Tortoises are basically reptiles with fused, expanded ribs. The ribs carry bony plates and a tough horn shell. Legs, head, and neck — protected by scaly skin — stick out from the box, or can be pulled in as the tortoise sees fit. When the tortoise senses danger and pulls them in, there is little for a predator to do but wait outside.

Though bony on the outside, this desert tortoise is a package of meat and internal organs that any predator would welcome as a meal. Perhaps more important, it contains in its living tissues more water than the thirsty fox could find for miles around. The fox may be patient, but it is hungry and cannot afford to hang around for more than a few minutes. The tortoise, parked in a patch of shade, can wait for hours. For an animal whose history spans 200 million years, what's the hurry?

Termites BLIND SOLDIERS

Termites, sometimes called "white ants" and often confused with ants, are more like beetles. They are similar to ants only in their community nesting. There are more than 1,500 species, living in warm temperate and tropical climates. Some live in timber, which they excavate and tunnel through. If you live in a tropical climate and your house falls down around your ears, you may well have been host to wood-boring termites.

Some kinds of termites build delicate, paper-thin nests that hang like Chinese lanterns from trees. Others build massive mounds of hardened soil and droppings that are up to 20 feet (6 m) tall, riddled with tunnels and chambers, and contain over a million busy termites. What are they doing? In the middle sit the king and queen termite, surrounded by helpers and feeders. The queen, 2 inches (5 cm) long with a hugely distended abdomen, lays several hundred eggs per day, which the king has fertilized, and the helpers tend. Worker termites emerge from the nest to scour the surrounding countryside for food, mostly rotting vegetation, to be carried back to the nest and turned into compost for the community. Soldier termites, in some species blind and guided only by scent, guard the nest from marauders. Here, two soldier termites stand guard at one of the entrances to the mound, eyeless but ready to attack ants or other insects that try to invade.

A soldier termite drawn seven or eight times life size.

Musk oxen WALL AGAINST WOLVES

Midway in size between cattle and sheep, musk oxen browse in small herds of a dozen or more on the tundra plains of Greenland and Canada. They are covered with dense woolly fur, which makes them look almost twice their real size. For defense and sparring, they have curved, sharply pointed horns.

In summer, when long daylight hours stimulate growth in the sparse vegetation, musk oxen live well and grow fat. In winter, food plants lie deep under hard-packed snow, which the oxen scrape away with their hooves. This is the hungry time, when they survive mainly on the fat that they stored during the summer.

Others, too, are hungry, including bands of timber wolves, which sometimes follow the musk oxen. In spring, when the calves are born, the wolves become particularly menacing. The musk oxen form a defensive ring, with calves and young animals in the middle. Even a dozen wolves attacking together stand little chance against that circle of lowered heads and sturdy horns.

Clown fish SAFETY IN DANGER

Coral reefs are busy underwater cities, where hundreds of species compete for living space and food. Competition is keen, predation fierce, and danger always imminent. For every morsel of food there are a dozen contenders, for every cavity or safe place a dozen home-seekers. Here is a fish that solves these problems in an unusual way. It makes a home among the stinging tentacles of a sea anemone — normally one of the reef's most dangerous places.

Clown fish, 3–4 inches (7.5–10 cm) long, dart among the corals and seaweeds, snapping up smaller fish and crustaceans. If startled, they seek the wall of the reef, which is often lined with clusters of sea anemones. These soft-bodied creatures open like flowers, to display waving tentacles covered with stinging cells. Animals that touch the tentacles are paralyzed and are drawn into the anemone, to be killed and digested.

Each clown fish seems to form a relationship with one particular anemone that, after a short "learning" period, accepts it and does not sting it. The fish can dart safely into its own anemone, where it is protected from predators that might try to follow. The anemone becomes its home, where it can raise a family of little clowns in complete safety. What does the anemone gain? Perhaps scraps of food or droppings from the fish, perhaps nothing — we do not know.

Like circus clowns, clown fish are brightly colored.

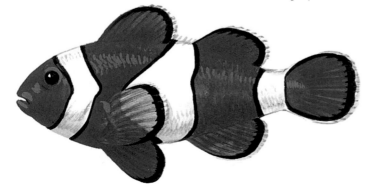

Five-lined skink CAMOUFLAGE AND DISPLAY

Animals that are liable to be attacked by predators — and very few are not — usually find safety in camouflage, being colored to match their background. If so, the last thing you might think they need is a brilliant patch of contrasting color to give them away. Here is an animal that seems to have changed its mind, or its strategy, halfway along. It is a North American five-lined skink, a reptile with short legs and a long, snakelike body and tail. Most of the body is sandy-brown with dark and pale stripes, in tones that match the desert soils around it. The tail by contrast is bright blue, matching nothing and sticking out like a sore thumb.

Skinks are particularly vulnerable to predation by birds. The reason for such a tail becomes apparent when its owner is under attack. This American crow has swooped down and grabbed at the skink, which has countered by splitting off its tail — there is a special weak point along the bone that allows this. The skink trots quietly off in one direction, the bright blue tail bounds and twists away in another. Not surprisingly, the crow follows the tail and swallows it down. Within a few weeks the skink has grown a new tail. So everyone gets something.

A five-lined skink
with tail intact.

Hermit crab NATURAL CAMOUFLAGE

Hermit crabs live in the discarded shells of whelks and other mollusks. Like other crabs, they start life as eggs, laid close to the seabed, then turn into larvae and float up into surface waters. The larvae feed on tiny plant cells in the plankton, change shape several times, and eventually return to the seabed as tiny baby crabs. Each young hermit crab finds a tiny shell and fits into it. As it grows, it graduates in stages to bigger shells. This soldier crab of the North Sea coast lives in a large "sentry box." If you picked it up for a close examination, its first line of defense would be to pinch you with its claws. But you might see how neatly its body twists to fit the shell spiral.

Crabs that recycle old mollusk shells do not need to grow shells of their own. They have the further advantage of looking like tough old whelks, rather than more vulnerable crabs. If the shells are decorated with sea anemones, they are harder still to spot on sea floors cluttered with those creatures. Many hermit crabs become encrusted in this way. The crabs gain protection and the anemones probably benefit from being carried around. They may even catch snacks floating up from meals served below.

Western tanager NO PEACE FOR OWLS

Tanagers are small, brightly colored songbirds, 6–7 inches (15–17 cm) long, which feed and nest in woodlands. Western tanagers, with brilliant yellow body, red head, and black wings, breed in western North America. Active during the day, they feed on fruit and insects, and roost by night among the woodland foliage.

Great horned owls are large, predatory birds, up to 2 feet (60 cm) long, with a wingspan of more than 6 feet (1.8 m), and distinctive gray — or brown — barred plumage. The "horns" are tufts of feathers on either side of the head. Widespread in both North and South America, they hunt by night, eating mainly insects and small mammals. During the day they roost, often close to the trunk of a tree where they are almost invisible.

A tanager going about its business in daylight has little to fear from a great horned owl. However, like many other small birds, in groups tanagers have a strong reaction to owls of any kind. The very sight of one brings on a kind of behavior similar to mobbing (see pages 16–17). If you see several tanagers darting and chirping excitedly around a particular tree, almost certainly they have found an owl and are drawing attention to it. The owl may then move on to a quieter spot, where it can sleep in peace.

Lantern fly WING-FLASHER

Here is a strange-looking South American insect that lives among trees and shrubs. It grows up to 3 inches (7.5 cm) long from its tail to the end of that curious "trunk," or nose. Lantern flies are pleasantly mottled in green and brown, to match a range of plant backgrounds. Matching is what it does best — that is its main defense against predators.

But touch a lantern fly lightly, and it will spread its wings just enough to reveal two spots that look remarkably — even frighteningly — like the eyes of a mammal or bird. Experimenters using different patterns of crosses, dots, and rings have shown that, of all symbols, birds are most frightened by the sudden appearance of "eyes" — a point that lantern flies have long relied upon.

Why are they called lantern flies? Several kinds of insects, such as fireflies, have the ability to produce intense light, from chemicals that they bring together in specialized body cells. Fireflies keep these cells in their tails and flash as they are flying, perhaps to attract each other or to repel attackers. Some lantern flies may be able to do the same, flashing from that long nose. This particular species does not. It relies entirely on flashing the eye-spots under its wing cases.

Sooty albatross SCARING OFF A SKUA

Sooty albatrosses are brown birds of the subantarctic regions. They nest on remote islands in the cold southern oceans, and feed on surface-living plankton and fish. Adults breed in spring, building a nest of grass and mud on a cliff ledge, where landing and takeoff are easy for them. Each pair lays a single white egg, taking turns to incubate it for a period of six or seven weeks.

The chick that emerges is cared for and fed by the parents alternately for the first three or four weeks. It grows rapidly, and at six to seven weeks can sit up in the nest unsupported, looking like a cuddly gray toy. Now both parents leave it, returning every few days with food in their crops, which the chick takes directly from their throats.

Between visits, the chick sits alone, sleeping or watching the world go by, occasionally standing to stretch its leg muscles, and flexing its rapidly growing wings. Even half-grown, however, it seems sadly vulnerable to any predator that comes along.

But not so. The most likely predators on the southern islands are brown skuas — powerful gull-like birds with a taste for eggs and chicks. If a skua lands close by, the sooty albatross chick turns like a weathervane to face it. If the skua approaches, the chick snaps its bill in an unmistakable warning. If the skua attacks, the chick gives a final snap, gulps, and throws up a jet of bad-smelling oil from its crop — the final remnants of its last meal. The jet, like a miniature fire hose, can be very accurate. It is enough to scare off the attacker. If it hits squarely, it will make the skua sticky and uncomfortable for several days.

Skunk CHEMICAL WARFARE

Spotted skunks, widespread across North America, are among the smallest of their kind, up to 22 inches (56 cm) long, with black-and-white checkerboard patterns on their flanks. They hunt by night in forests or along the woodland edge. You are never in doubt if they are around. You hear their asthmatic sniffing and grunting, and may see white flashes in the undergrowth. But above all, you smell them — a strong, sharp, horrible smell that carries far on the wind.

Skunks feed on insects, birds, and small mammals. Left to themselves, they mind their own business, hunting and raising stinky little families. Despite their smell, they are hunted and eaten by some of the larger owls and cats. Skunks have sharp teeth and claws, but their main defense lies at their other end. The source of their smell is two glands under the tail, which secrete an acrid, blinding and very smelly fluid. They can eject it in jets up to 10 feet (3 m).

Threatened by a predator, like this young cougar, the skunk turns tail and raises his hind legs in the air. At this point, an experienced predator retreats. This young cat, about to get the full blast, will quickly learn the cost of annoying a spotted skunk.

African porcupine SPIKES FOR DEFENSE

Porcupines are rodents, relatives of rats and squirrels, with large front teeth for gnawing and biting. They live in the Americas, Africa, and Asia, differing slightly according to their origins. American porcupines, for example, are tree-climbers, while those of Asia and Africa are almost entirely ground-dwelling.

Wherever they live, porcupines have a number of sharp, stiff spines called quills in the fur of their backs. These are modified and strengthened hairs, backward-pointing, and capable of being raised and lowered. Porcupines feed mainly on fruit, shoots, and bark, but they sharpen their teeth on a range of materials including telegraph posts and automobile tires.

This African brush-tailed porcupine, about 24 inches (60 cm) long, is slightly smaller than its American cousins. A young predator such as a leopard, meeting one for the first time, has a surprise in store. The porcupine's first response to a strange animal is to turn away and present its backside. Then it raises its quills and rattles them. Finally, if still alarmed, it will rush backward, driving some of its longer quills into the enemy. At the very least, this will be painful enough to remind the predator never again to mess with a porcupine. But it might be more serious, even fatal, for the spines are difficult to extract, and sometimes cause long-term festering wounds.

Virginia opossum PLAYING DEAD

The expression "playing possum" means pretending to be dead. Here is the animal that gave its name to that saying. Virginia opossums are marsupials, more closely related to kangaroos and wombats than to the rats or marmosets they resemble. They share the marsupial characteristic of producing their young at a very immature stage, only about two weeks after conception, and keeping them in a pouch or traveling crib on their abdomen. Permanently attached to nipples, the babies feed on demand. They grow quickly to an intermediate stage where they can leave the pouch and travel on mother's back. A litter of a dozen or more growing opossums becomes a considerable load. But at about 15 weeks the babies reach independence.

An opossum that is attacked by a predator, such as this bobcat, will scratch and bite with the best. Then, if it seems to be losing, it suddenly goes limp and relaxes with eyes closed, as though dead. This may seem like an invitation to be eaten. However, for many hunters, including cats, killing and eating are separate activities. Once they have killed, they turn away from their prey until they feel ready to eat. And that, of course, is the opossum's chance to "wake up" and sneak away.

Fire-bellied toad BRIGHT WARNING

Though everyday toads of ponds and gardens rely for their safety on dull coloration, often they have a second line of defense. Glands in their warty skin secrete fluids that give them a nasty smell or taste. A predator that takes a toad in its mouth will spit it out quickly, and probably remember never to pick up another one. In effect, toads teach predators not to interfere with anything that looks toad-like.

To reinforce the lesson, fire-bellied toads make themselves even more memorable. Their first line of defense is still protective coloration. Seen from above on the muddy floor of a pond or stream, particularly in murky water, they look exactly like their background. But if a predator like this Chinese pond heron disturbs one, perhaps by accident, and is about to eat it, the toad brings into action a remarkable Plan B.

Sitting up on its hind legs, the toad stretches and falls backward, exposing a brilliantly colored yellow, orange, or crimson underbelly. At the same time, it secretes a white fluid from its skin, with a strong smell and foul taste. The heron will at least be astonished, and, however absentminded, is unlikely to forget the message — fire-bellied toads are frightening, trouble, horrible to taste, and best left alone.

Index CREATURES AND FEATURES